OnBoard
ACADEMICS

I0067176

Area, Perimeter and Volume
(4-6)

© 2015 OnBoard Academics, Inc
Portsmouth, NH
800-596-3175
www.onboardacademics.com
ISBN: 978-1-63096-089-6

OnBoard Academic's books are specifically designed to be used as printed workbooks or as on-screen instruction. Each page offers focused exercises and students quickly master topics with enough proficiency to move on to the next level.

OnBoard Academic's lessons are used in over 25,000 classrooms to rave reviews. Our lessons are aligned to the most recent governmental standards and are updated from time to time as standards change. Correlation documents are located on our website. Our lessons are created, edited and evaluated by educators to ensure top quality and real life success.

Interactive lessons for digital whiteboards, mobile devices, and PCs are available at www.onboardacademics.com. These interactive lessons make great additions to our books.

You can always reach us at customerservice@onboardacademics.com.

Solid Figures

Key Vocabulary

solid figure

base

vertex

edge

face

net

Draw a line to connect the shape with its name.

Sphere   **Rectangular Prism**   **Cube**

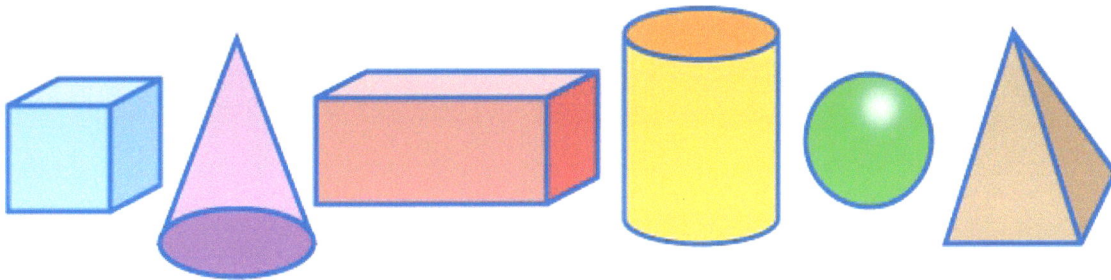

Pyramid   **Cylinder**   **Cone**

Prisms

This prism has:

face ●——→

edge ●——→

vertex ●——→

base ●——→

[ ] faces

[ ] edges

[ ] vertices

[ ] bases

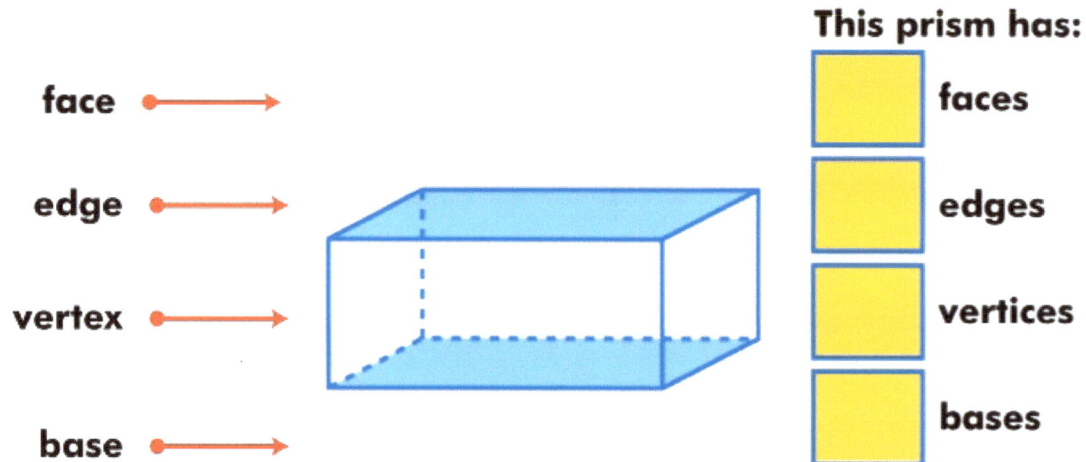

Label one face, one vertex, one edge, and one base of this prism.

A prism has two parallel congruent bases. It is named by the shape of its base. The other faces are always rectangles.

Pyramids

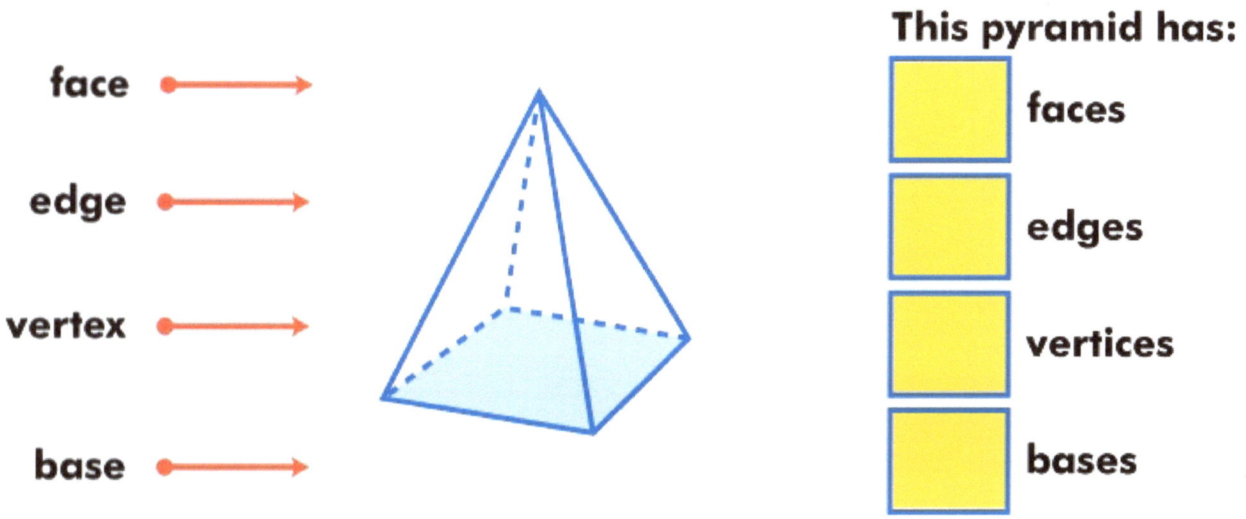

face •——————→

edge •——————→

vertex •——————→

base •——————→

This pyramid has:

faces

edges

vertices

bases

Label one face, one vertex, one edge, and the base of this pyramid.

A pyramid has one base. It is named by the shape of its base. The other faces are always triangles.

Three solids with curved surfaces

Label the base(s) and curved surface(s) of each solid

Cylinder

Cone

Sphere

curved surface ●———▶

base ●———▶

Label the illustrations below

| Prism | or | Pyramid |

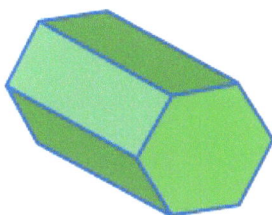
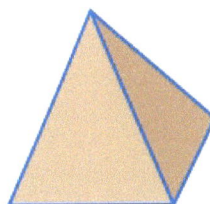

| One base | | Two parallel bases |

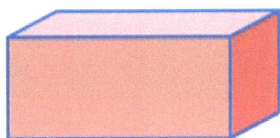
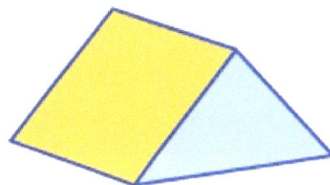

The unfolded box

a box

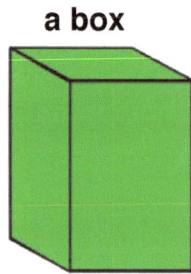

an unfolded box

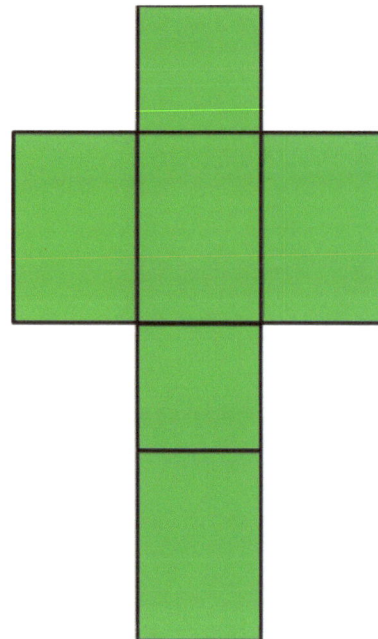

This is the net of a rectangular prism.
A net is a flat diagram that can be
folded to create a solid figure.

Match the solid with its unfolded version.

Enter the number of the yellow unfolded solid in the box below the corresponding blue solid.

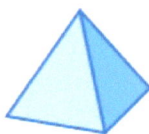 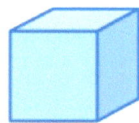  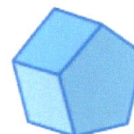

Square Pyramid	Cube	Triangular Prism	Pentagonal Prism	Tetrahedron
☐	☐	☐	☐	☐

  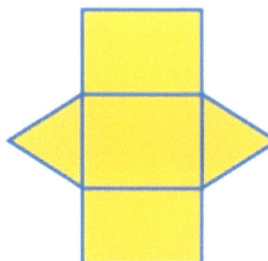  

①   ②   ③   ④   ⑤

Name: _____

Solid Figures Quiz

1 True or false? A prism has two parallel congruent bases. The other faces are always triangles.

2 This solid has five faces, five vertices, and eight edges. Four of the faces are triangles.

   A Rectangular prism

   B Triangular prism

   C Square pyramid

   D Pentagonal pyramid

3 How many faces does a triangular prism have?

4 How many edges does a cube have?

Percents

Key Vocabulary

percent

fractional percent

ratio

Percent means "per hundred."

25% of this grid is shaded blue

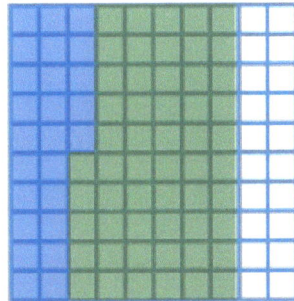

What is the value of the green shade?

What percent of the grid is not shaded?

Find the percents.
When possible write the percent as a decimal and a fraction.

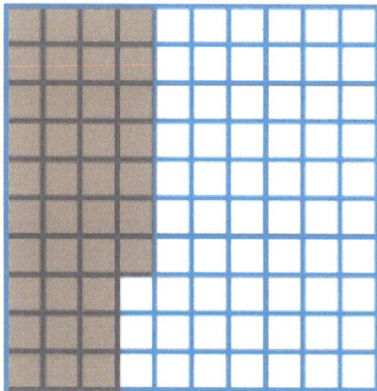 

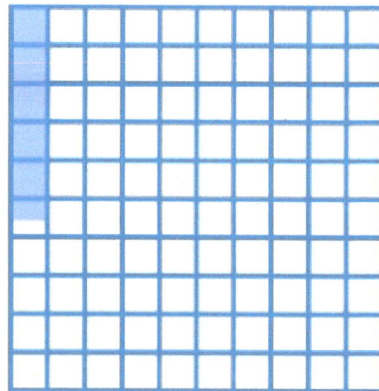 

or

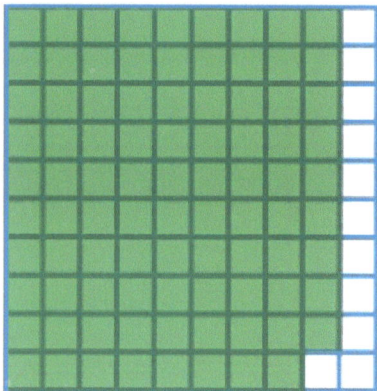 

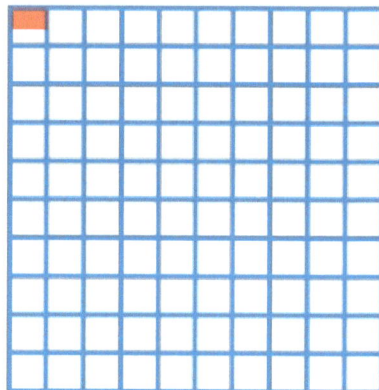 

or

Practice percents with this vacation survey.

In a survey, 100 people were asked to name their ideal vacation. The results are shown in this 100 grid.

%

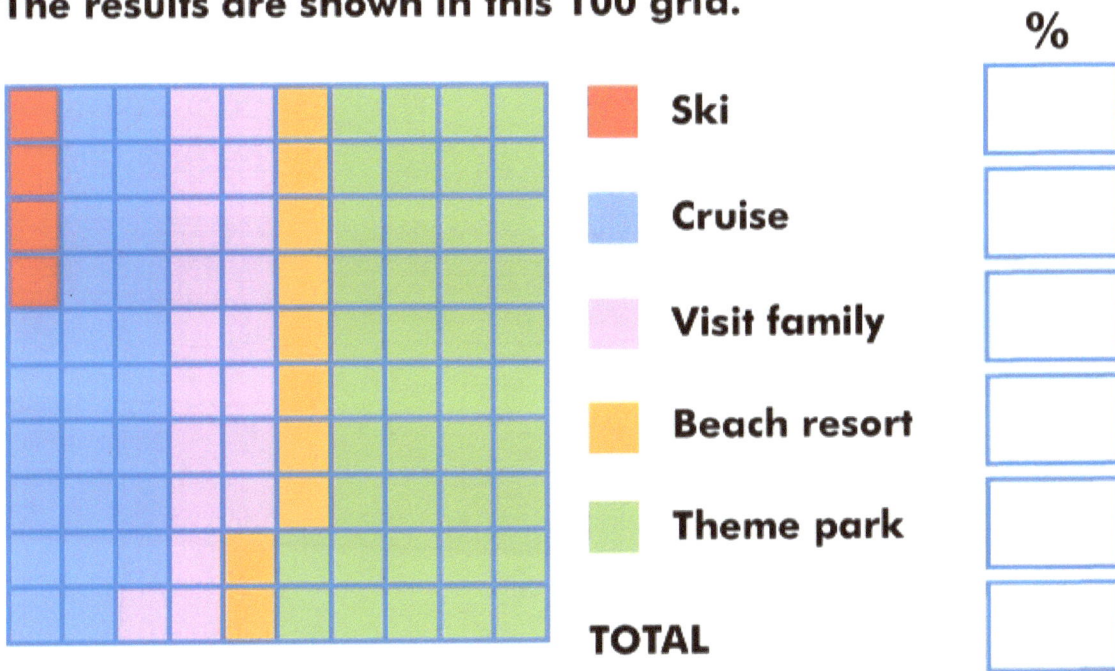

🟥 Ski	
🟦 Cruise	
🟪 Visit family	
🟧 Beach resort	
🟩 Theme park	
TOTAL	

What percent of the people surveyed chose each vacation?

Thanksgiving Lunch!
Complete the table.

	To:	All staff and students
Send	Subject:	Thanksgiving lunch

Hi everyone,

I'm going to cook a special Thanksgiving lunch and I'd like to know what you would all like. Email me and I'll total up the votes, and we'll go with the most popular choice!

Best Wishes, Ron

Thanksgiving lunch selection	Students	%	Staff	%
🍗	29		16	
🍕	13		3	
🍔	20		6	
🌭	38		0	
TOTAL	100	100	25	100

$$\frac{16}{25} = \frac{?}{100}$$

$$\frac{3}{25} = \frac{?}{100}$$

$$\frac{6}{25} = \frac{?}{100}$$

$$\frac{0}{25} = \frac{?}{100}$$

Find the missing values.

Survey Results	Ratio	Out of 100	Percent
3 out of 10 people had visited Europe.	——	——	
24 out of 25 trains ran late yesterday.	——	——	
4 out of 5 cars in the parking lot were SUVs.	——	——	
15 out of 20 students said they liked math.	——	——	

What is Mayor Lewis' approval rating?
Another method for finding percent from a ratio.

33 out of 40 people interviewed gave Mayor Lewis a positive approval rating. Express Mayor Lewis' approval rating as a percent.

1. **Write as ratio** $\dfrac{33}{40}$

2. **Divide numerator by denominator**

$$40\overline{)33.000} \quad \text{(quotient } 0.825\text{)}$$

3. **Multiply by 100%** $\quad 0.825 \times 100\% = 82.5\%$

More about Mayor Lewis.
See if you can figure out how many people approve of the Mayor's education policies. If they approve of the policy, they will vote for him in the next election.

Only 23 out of 40 people surveyed, approved of Mayor Lewis' education policies.

What percent approved of the Mayor's education policies?

What percent disapproved?*

What percent will vote for the mayor at the next election?

*****Figures assume those who didn't approve disapproved.**

Name_____

Percent Quiz

1 True or false? 94.5% of this grid is not shaded.

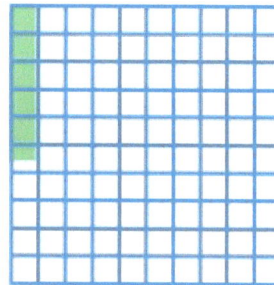

2 In a class of 25 students, 8 have a cell phone. Write this as a percent.

8% **A**   16% **B**   25% **C**   32% **D**

3 In a class of 25 students, 13 travel to school by bus. Write this as a percent (number only).

4 In a group of 40 athletes, 27 are female. Write this as a percent (number only).

Area

Key Vocabulary

area

Parallelogram

square unit

The area of this rectangle is 8 square inches.

To measure the parallelogram how would you make a rectangle?

Step 1

Step 2

Step 3

The formula for the area of a parallelogram.

Rectangles and parallelograms with the same base and the same height have the same area.

$$A = b \times h \qquad A = b \times h$$

Find the area for the yellow triangles.
The dotted lines provide a hint.

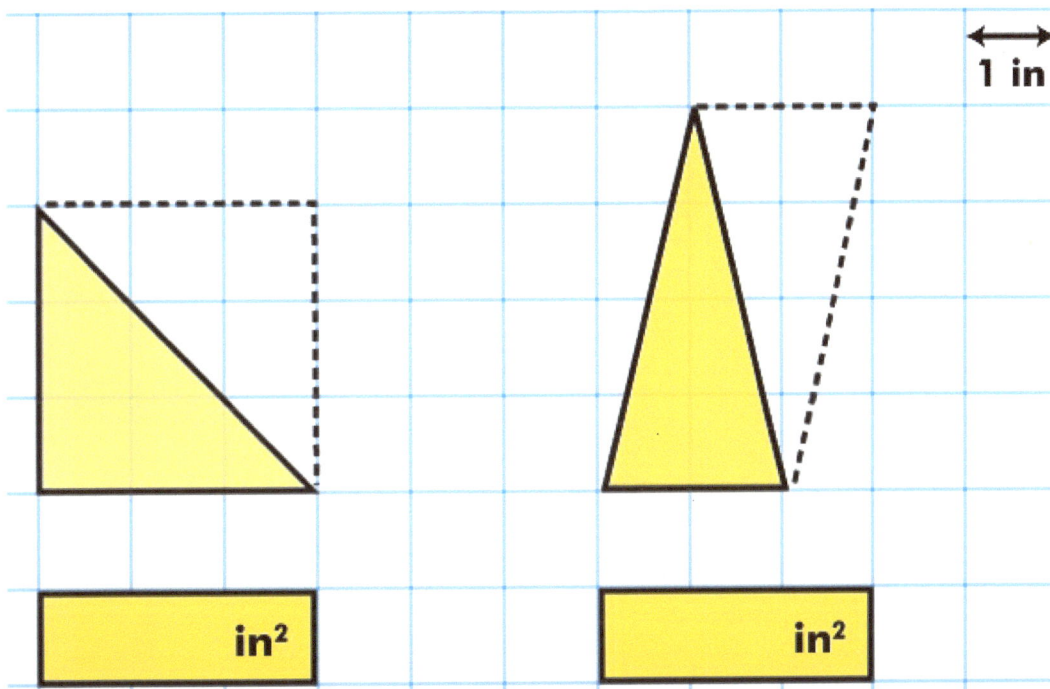

1 in

| in² | in² |

The area of a triangle is half the area of a parallelogram.

Triangles and half parallelograms
Study the illustration below to discover the the formula for the area of a triangle.

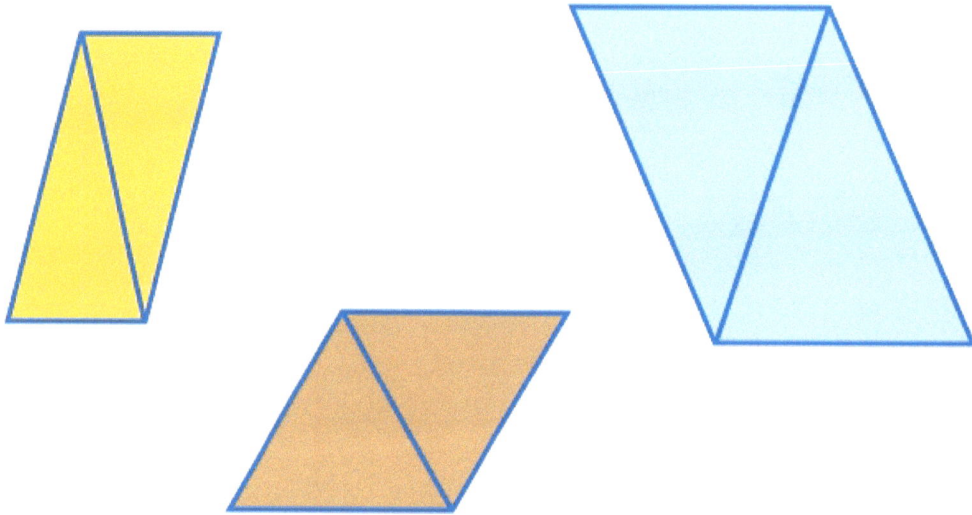

Area of a parallelogram = base x height

Area of a triangle = $\frac{1}{2}$(base x height)

Name: _____

Area Quiz

1 True or false? The area of a parallelogram is half the area of a triangle.

2 What is the area of Figure 1?

  A 7.2 cm²   **B** 4.8 cm²   **C** 7.8 cm²   **D** 6.2 cm²

3 What is the area of Figure 2 in sq ft?

4 What is the length of the base of Figure 3 in feet?

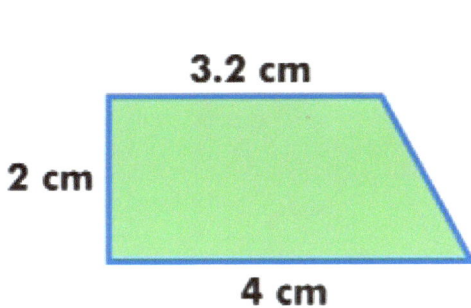

3.2 cm

2 cm

4 cm

FIGURE 1

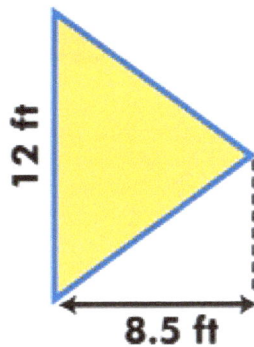

12 ft

8.5 ft

FIGURE 2

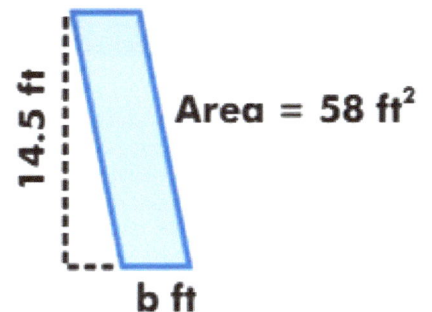

14.5 ft

Area = 58 ft²

b ft

FIGURE 3

Area

Key Vocabulary

area

Parallelogram

square unit

The area of this rectangle is 8 square inches.

1 in

8 in²

To measure the parallelogram how would you make a rectangle?

Step 1

Step 2

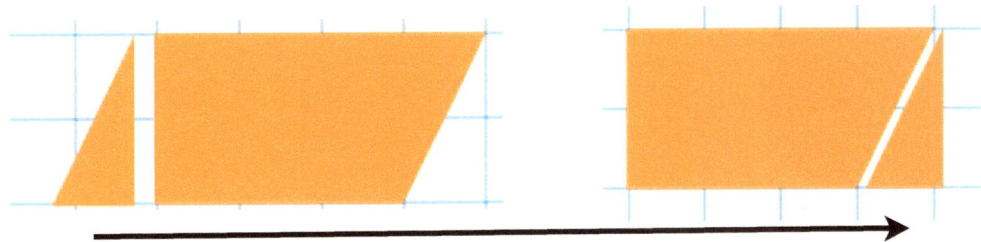

Step 3

The formula for the area of a parallelogram.

Rectangles and parallelograms with the same
base and the same height have the same area.

$$A = b \times h$$

$$A = b \times h$$

Find the area for the yellow triangles.
The dotted lines provide a hint.

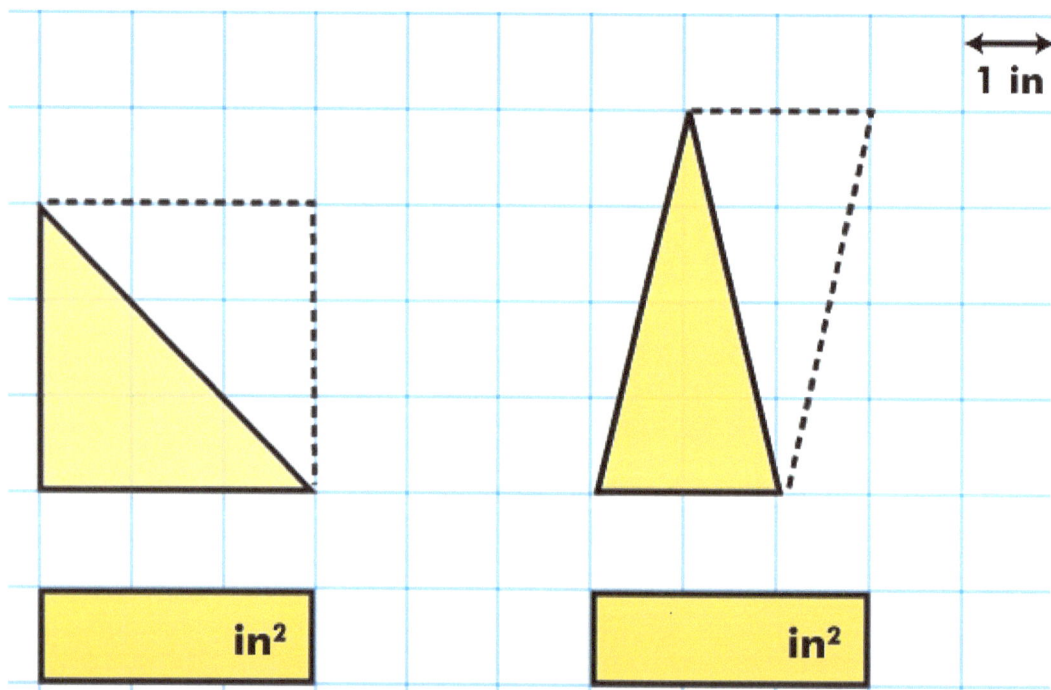

1 in

in²

in²

The area of a triangle is half the area of a parallelogram.

Triangles and half parallelograms
Study the illustration below to discover the the formula for the area of a triangle.

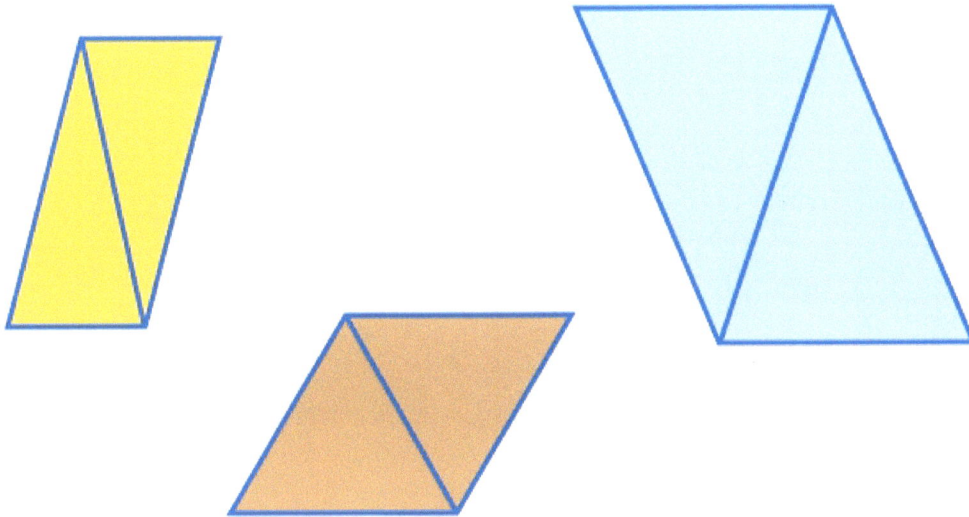

Area of a parallelogram = base x height

Area of a triangle = $\frac{1}{2}$ (base x height)

Name: _____

Area Quiz

1 True or false? The area of a parallelogram is half the area of a triangle.

2 What is the area of Figure 1?

   A 7.2 cm^2 **B** 4.8 cm^2 **C** 7.8 cm^2 **D** 6.2 cm^2

3 What is the area of Figure 2 in sq ft?

4 What is the length of the base of Figure 3 in feet?

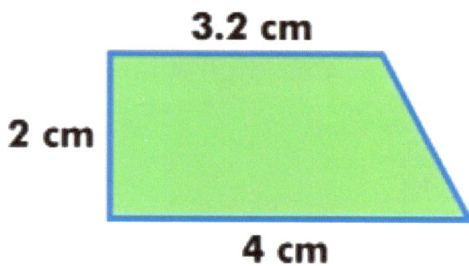

3.2 cm

2 cm

4 cm

FIGURE 1

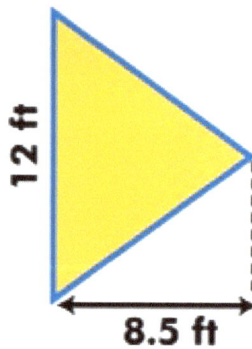

12 ft

8.5 ft

FIGURE 2

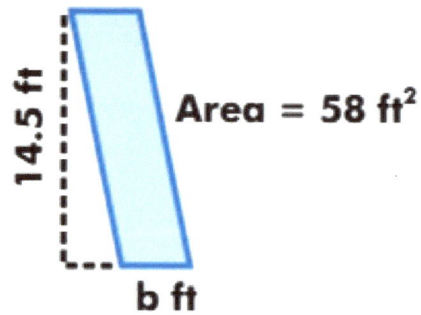

14.5 ft

Area = 58 ft^2

b ft

FIGURE 3

Area of Irregular Figures

Key Vocabulary

area

irregular figures

Study the illustrations below to discover strategies for finding the area of an irregular figure.

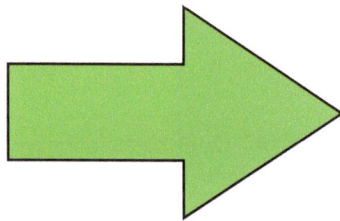

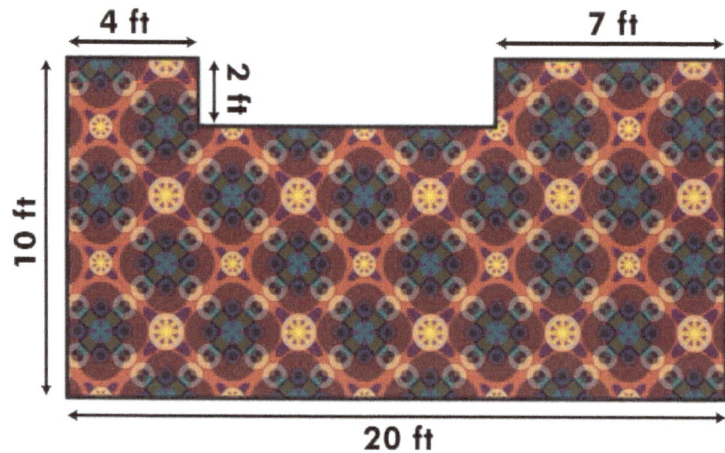

Ben's parents have reluctantly agreed to let him host a party, but they've insisted that he cover the carpet with a protective plastic sheet. What size sheet does he need?

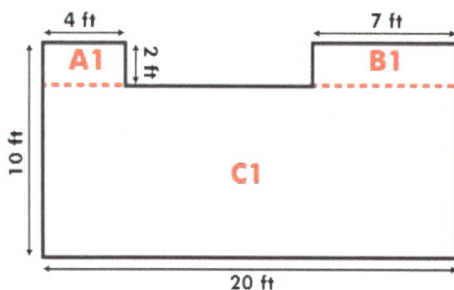

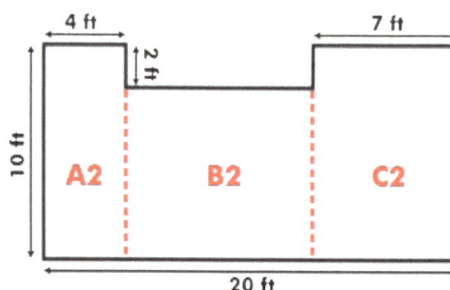

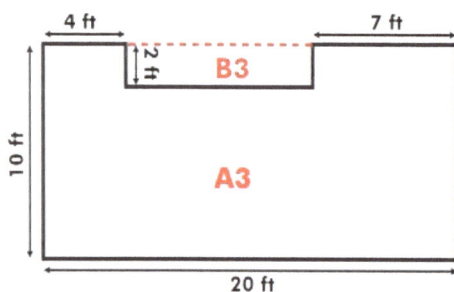

A1   B1   C1
(4 x 2) + (7 x 2) + (20 x 8) = 182

A2   B2   C2
(4 x 10) + (9 x 8) + (7 x 10) = 182

A3   B3
(20 x 10) − (9 x 2) = 182

3 Strategies to Solve

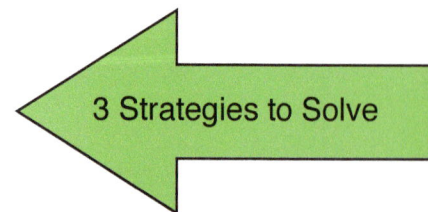

Ben needs 182 sq ft of sheeting to cover the carpet.

Practice dividing irregular shapes into familiar shapes.

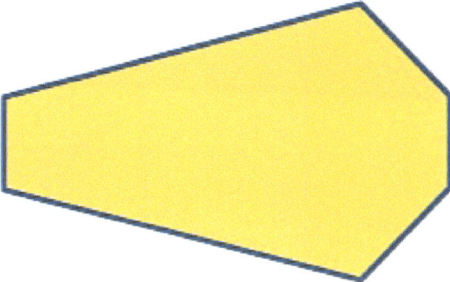

Find the area of these irregular shapes.

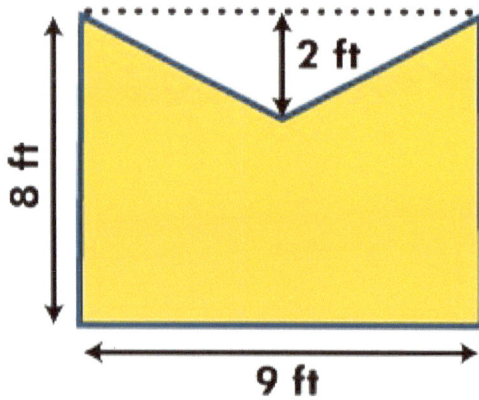

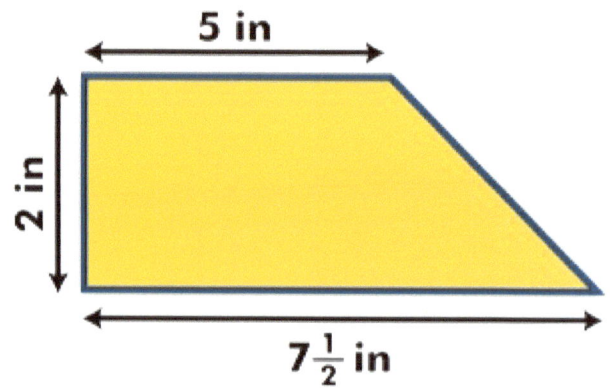

Find the area of this irregular hexagon.

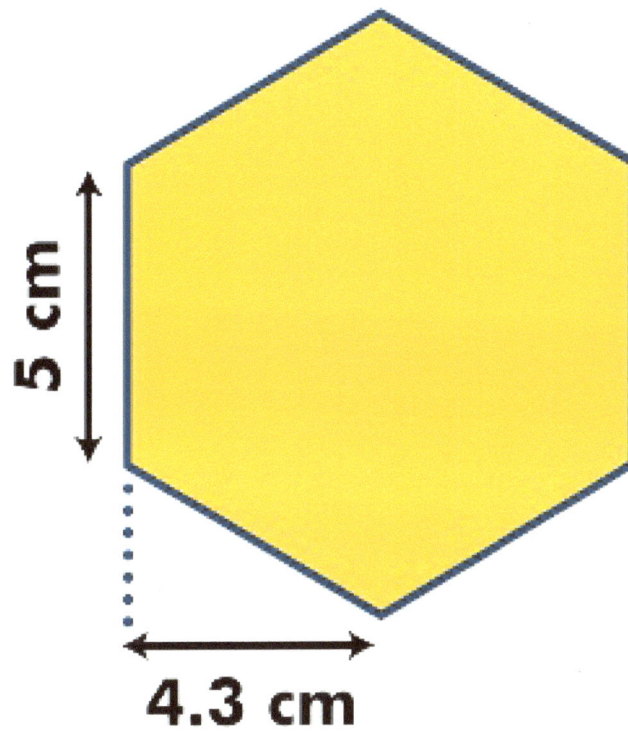

5 cm

4.3 cm

Who has the largest deck?

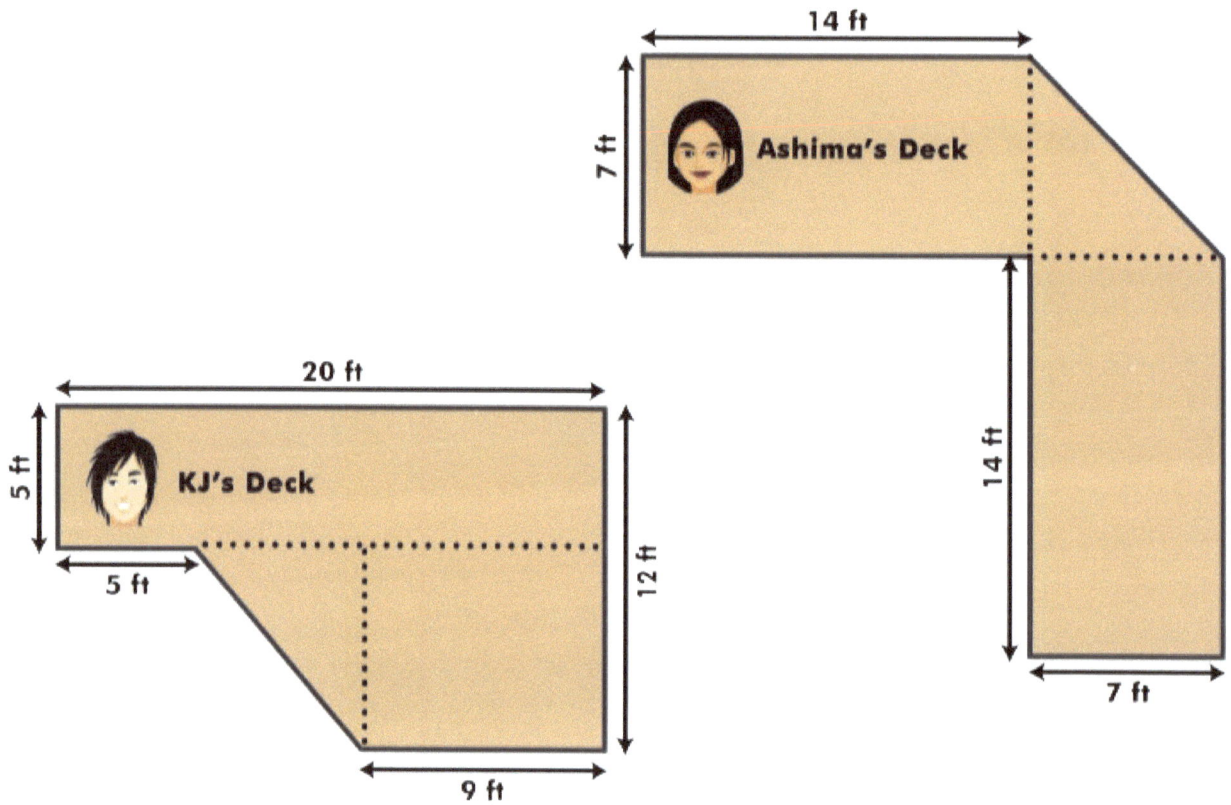

14 ft

Ashima's Deck

7 ft

14 ft

7 ft

20 ft

KJ's Deck

5 ft

5 ft

12 ft

9 ft

Name_____

Area of Irregular Figures Quiz

1 True or false? The area of the arrowhead (the triangle) in Figure 1 is 12 sq units.

2 The total area of Figure 1 is:

- **A**   38 sq units
- **B**   40 sq units
- **C**   44 sq units
- **D**   32 sq units

3 What is the area of Figure 2 in sq units?

4 What is the value of x in Figure 3?

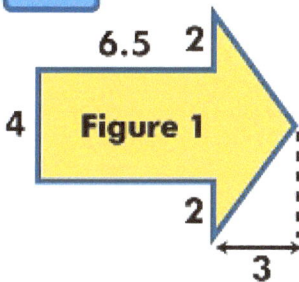

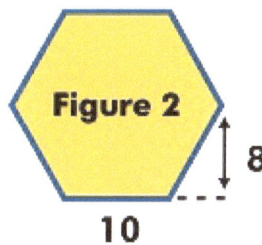

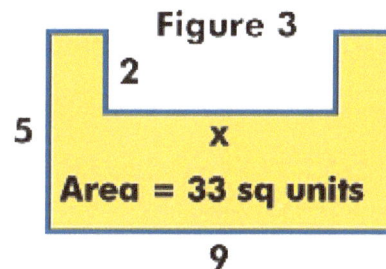

6.5   2

4   Figure 1

2

3

Figure 2

10

8

Figure 3

2

5

x

Area = 33 sq units

9

Perimeter and Area of Irregular Figures

Key Vocabulary

perimeter

area

square units

Find the perimeter of this room.

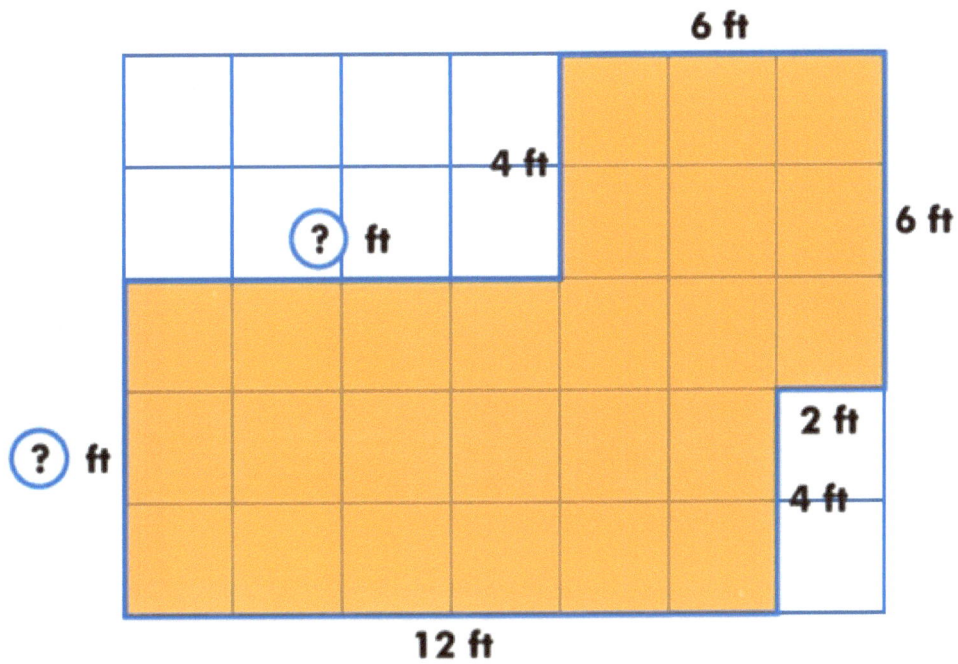

Find the area of this room.

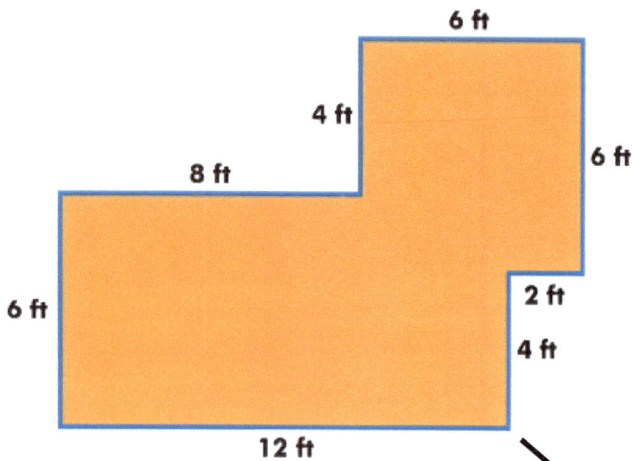

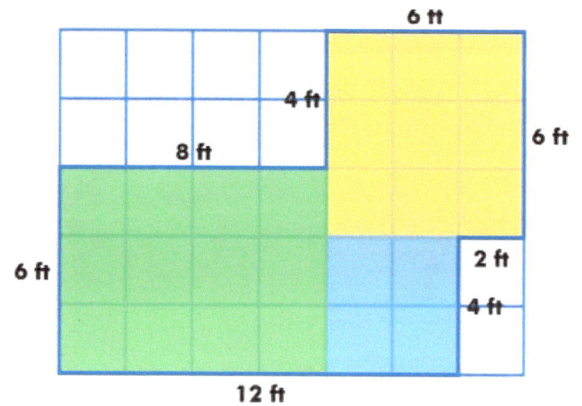

6 ft

4 ft

8 ft

6 ft

6 ft

2 ft

4 ft

12 ft

6 ft

4 ft

8 ft

6 ft

6 ft

2 ft

4 ft

12 ft

Area of green rectangle [ ] ft²   **Total area of room** [ ] ft²

Area of blue rectangle [ ] ft²

Area of yellow rectangle [ ] ft²

Find the perimeter and area for these shapes.

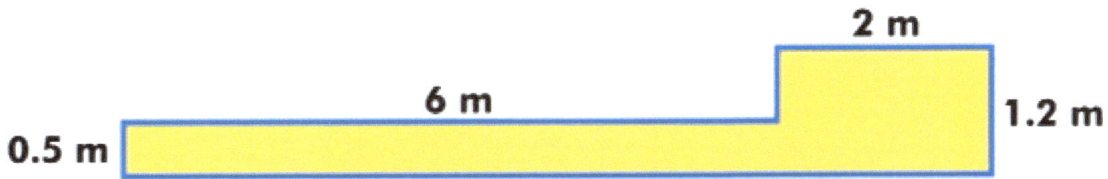

2 m

6 m

0.5 m

1.2 m

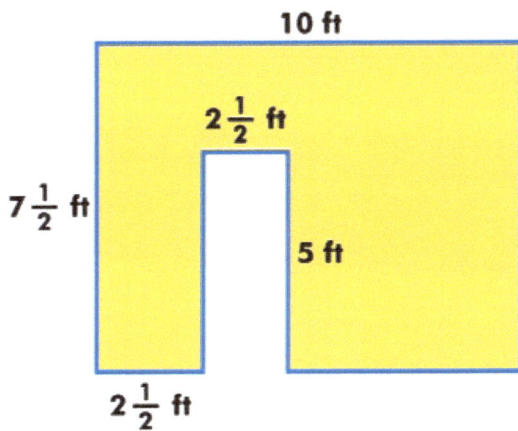

10 ft

$2\frac{1}{2}$ ft

$7\frac{1}{2}$ ft

5 ft

$2\frac{1}{2}$ ft

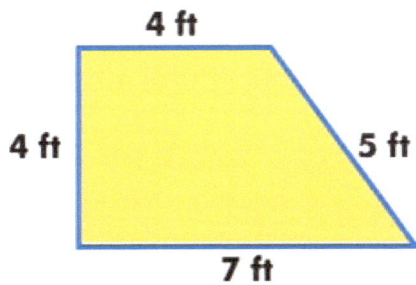

4 ft

4 ft

5 ft

7 ft

Find the perimeter and area of Mrs. Jones' apartment.

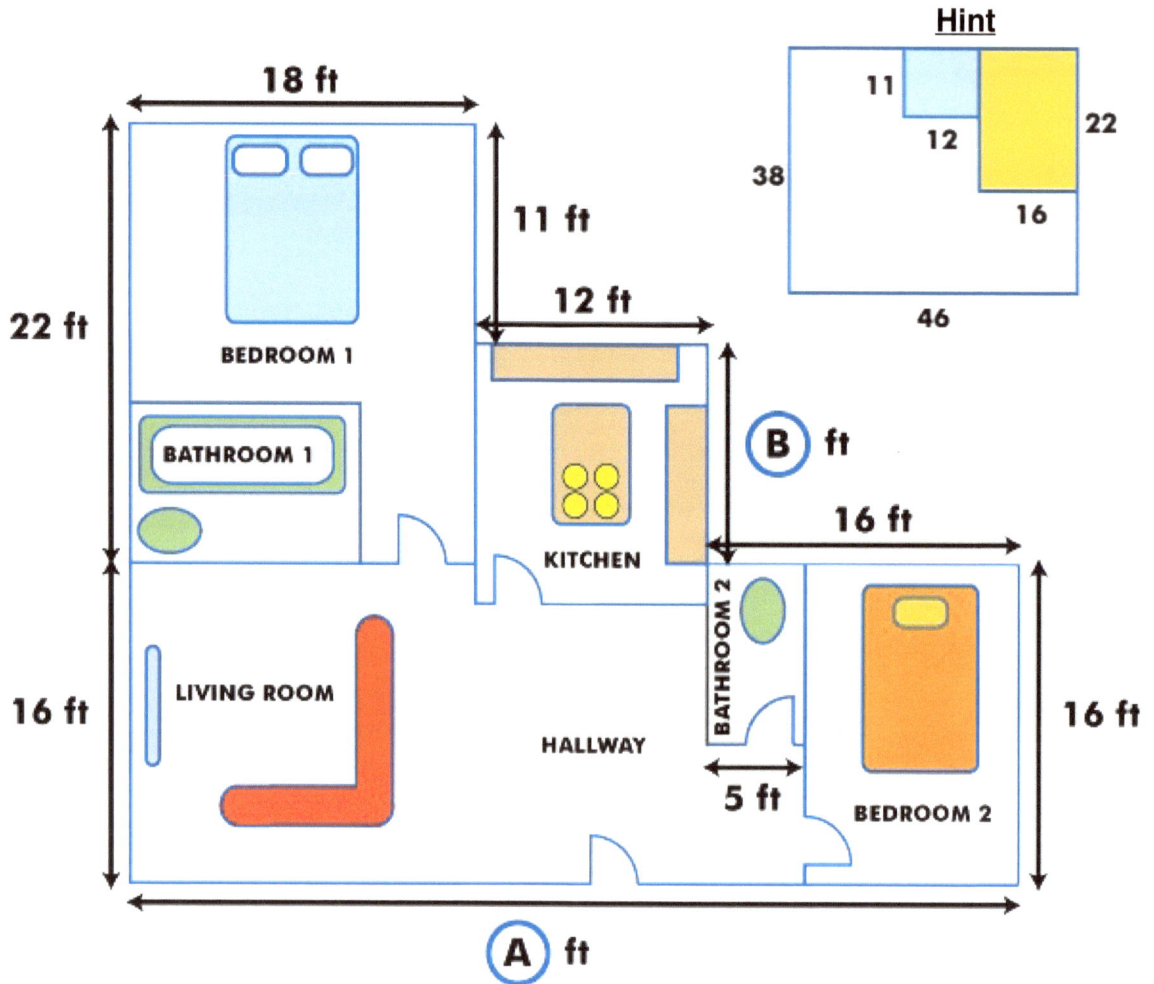

Hint

18 ft

11 ft

12 ft

22 ft

BEDROOM 1

BATHROOM 1

KITCHEN

B ft

16 ft

16 ft

LIVING ROOM

BATHROOM 2

HALLWAY

5 ft

BEDROOM 2

16 ft

A ft

11

12

22

38

16

46

Estimate the perimeter and area of this shape.

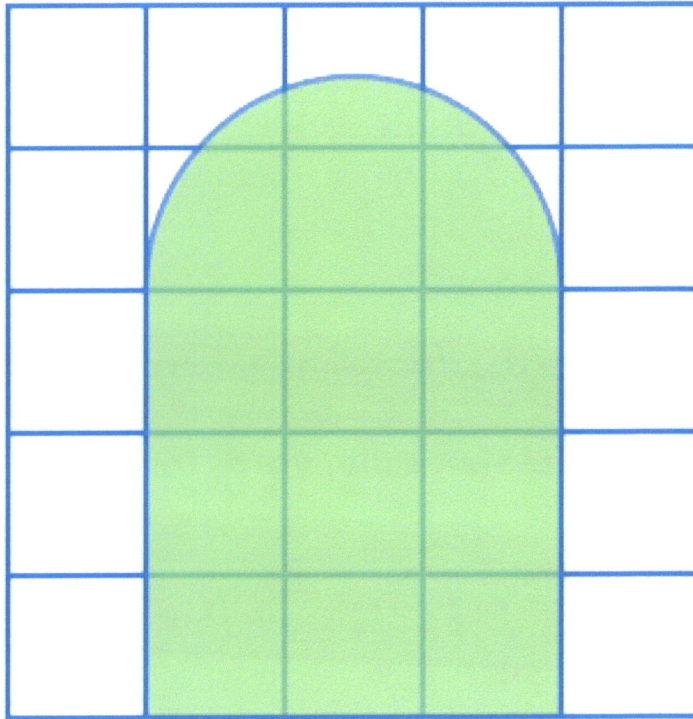

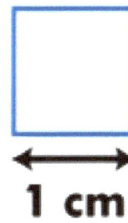

1 cm

Name: _____

Perimeter and Area of Irregular Figures Quiz

1 The perimeter of this rectangle is 2a + 2b. True or false?

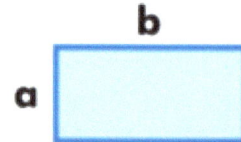

2 Which are all units of area?

    A ft^2, m^2, square yard, square mile, acre

    B feet, centimeter, meter, mile, yard, inch

    C ft^2, m^2, cm^3, mm^3

    D ft^3, m^3, cm^3, mm^3

3 The perimeter of this figure is 6a + ___ b

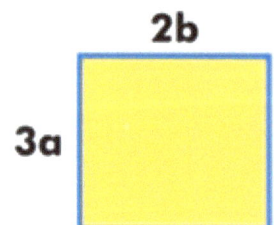

4 The area of this figure is ___ab

Volume of a Rectangular Prism

Key Vocabulary

volume

rectangular prism

unit cube

How many unit cubes are in a prism?_____

Unit Cube

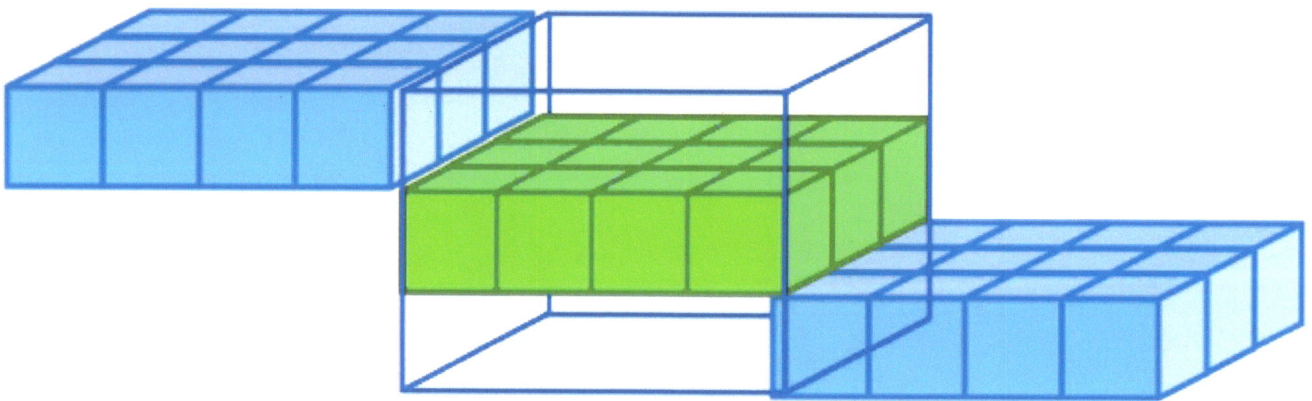

Do you need to count every cube?

_____

_____

Draw a prism using 27 cubes.

Draw a prism using 16 cubes.

How many cubes fit into this prism?

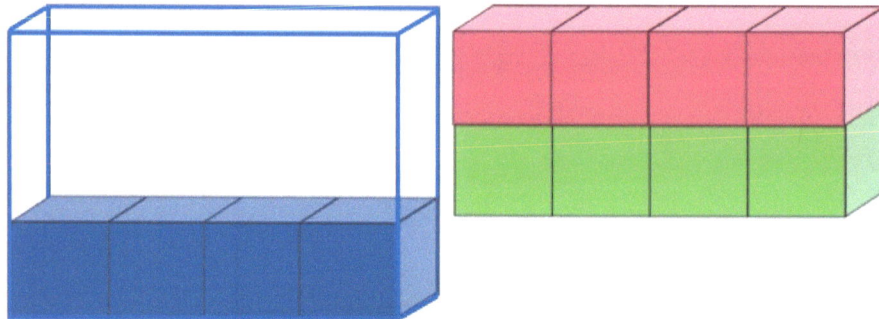

Unit cubes in first layer

Number of layers

How many cubes fit into this prism?

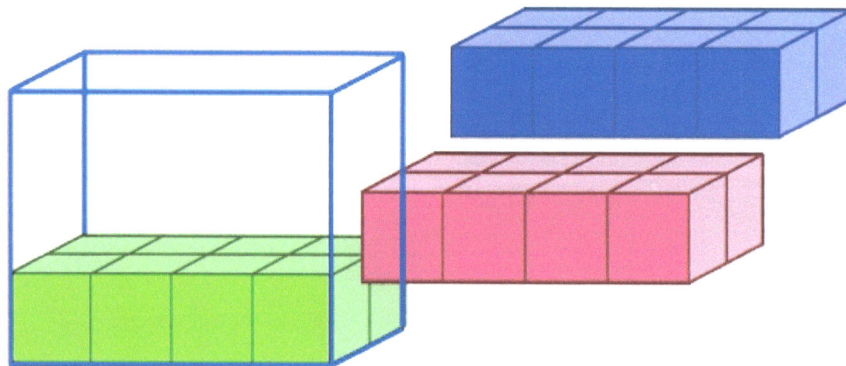

Unit cubes in first layer

Number of layers

How many cubes are there in a prism?

How many cubes in the first layer? ☐

How many layers? ☐

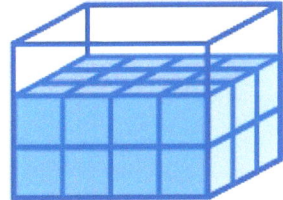

Total number of cubes? ☐ x ☐ = ☐

Volume of this rectangular prism? ☐ cubic units

Find the volumes of these prisms.

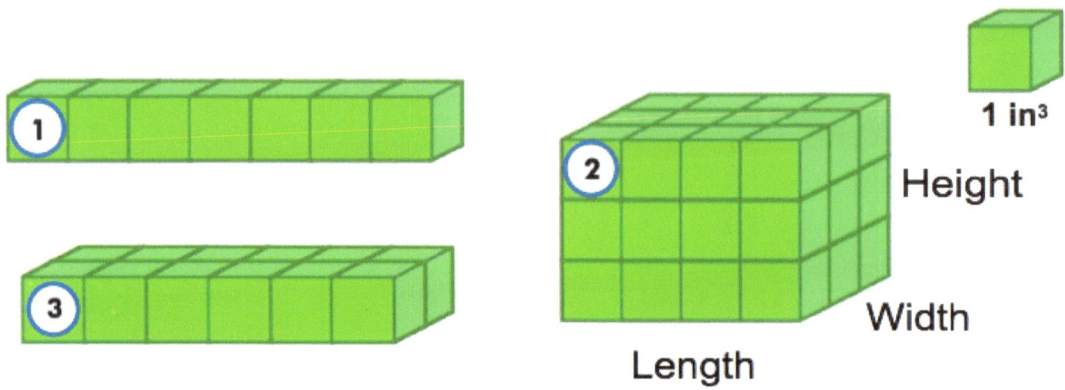

1 in³

Height

Width

Length

	Length	Width	Height	Volume
1	in	in	in	in³
2	in	in	in	in³
3	in	in	in	in³

Find the volume of these prisms.

1) _____

2) _____

3) _____

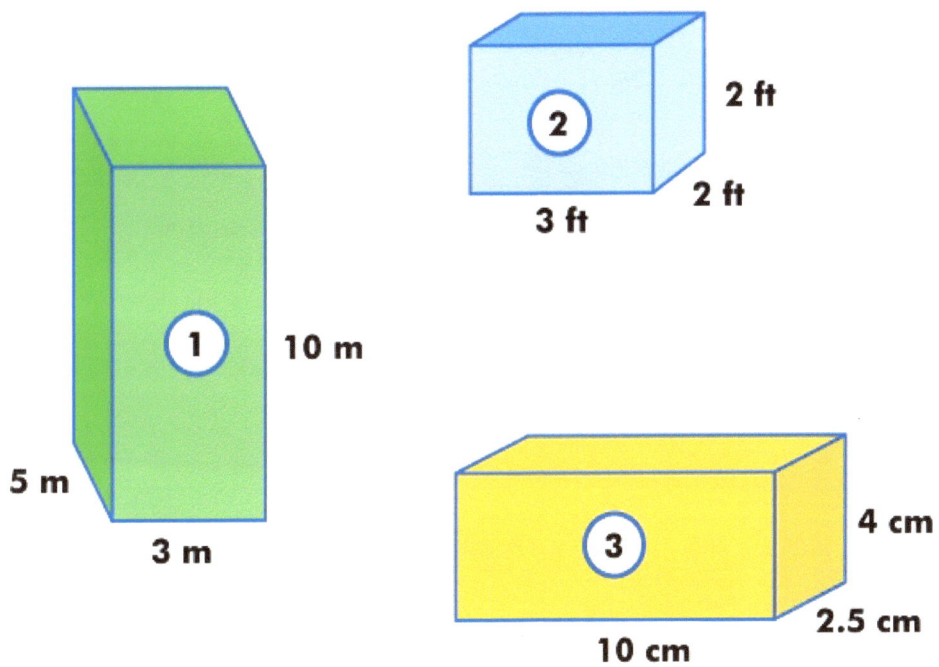

Find the missing dimension.

Alison's parents have purchased a swim training pool which holds 6,000 ft³ of water.

What is the depth of the pool?

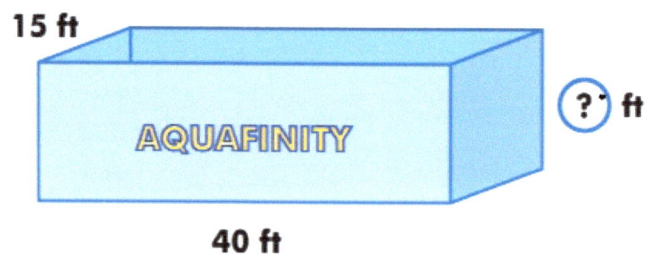

15 ft

AQUAFINITY

(?) ft

40 ft

Hint

$$40 \times 15 \times \boxed{?} = 6,000$$

Name: _____

Volume of a Rectangular Prism Quiz

1. True of false? Perimeter, area and volume can all be measured in units of square feet and square inches.
2. What is the formula for the volume of a rectangular prism?
    a. $(l \times w \ h)^3$
    b. l x w x h
    c. 2l x 2w x 2h
    d. l + w + h
3. What is the volume, in cubic units of this figure? _____

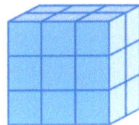

4. What is the volume, in cubic units, of this figure? _____

5. The volume of this prism is 24 units.

True

False

Nets and Surface Area

Key Vocabulary

surface area

net

prism

cube

rectangular prism

How many faces does a cube have?

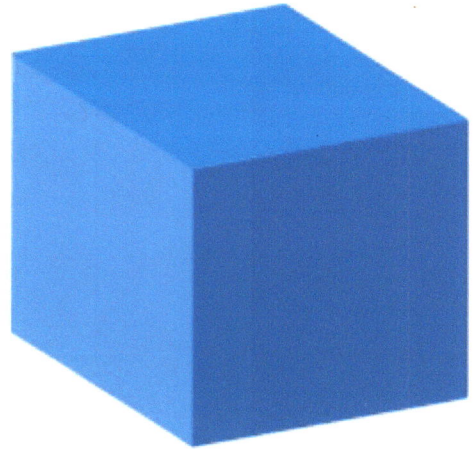

Take it apart to solve.

Finish the nets.

A net is a pattern that you can cut and fold to make a model of a solid shape. In this case we want to be able to make a cube. Complete the nets by shading in the additional squares to make a cube.

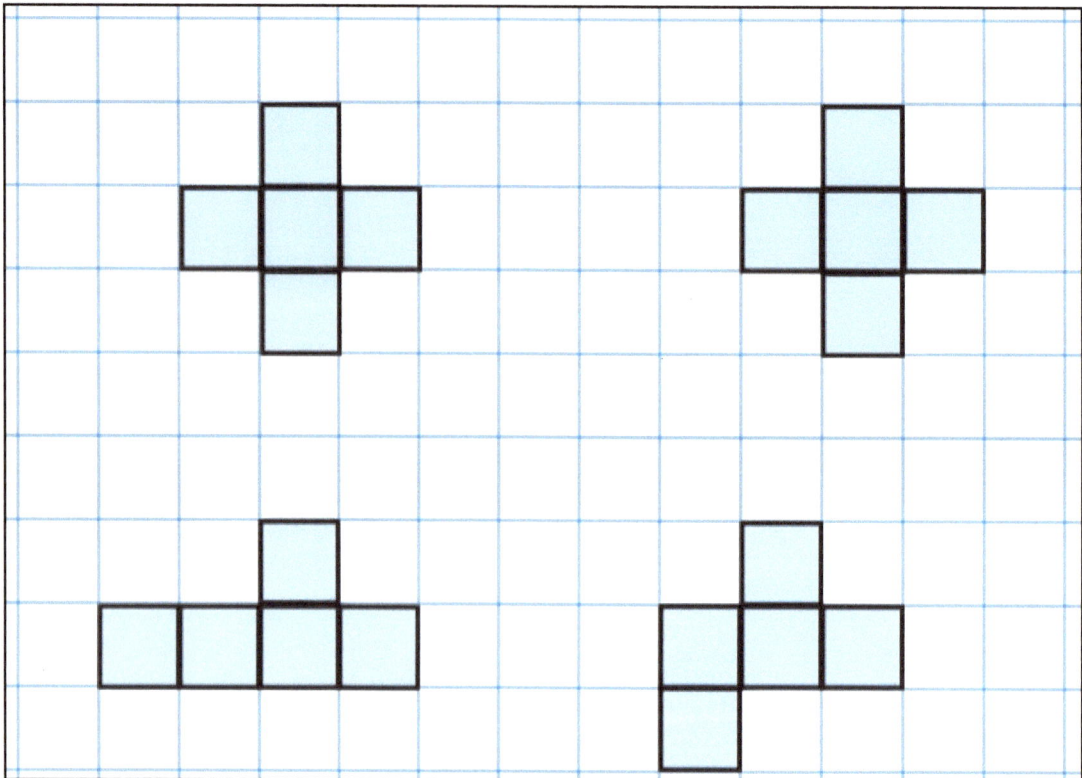

Which of these nets will make a cube.
Place a Y in the box for yes and an N for no.

The surface are of a cube.

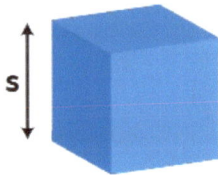

Area of one face: $s \cdot s = s^2$

Surface area of cube: $6s^2$

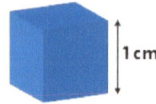

Area of one face: 1 cm x 1cm = 1 cm²
Area of six faces: 6 x 1 cm² = 6 cm²
Surface area = 6 cm²

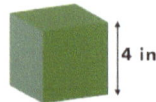

Area of one face: 4 in x 4 in = 16 in²
Area of six faces: 6 x 16 in² = 96 in²
Surface area = 96 in²

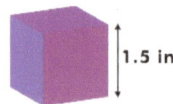

Area of one face: 1.5 in x 1.5 in = 2.25 in²
Area of six faces: 6 x 2.25 in² = 13.5 in²
Surface area = 13.5 in²

What is the surface are of these cubes?

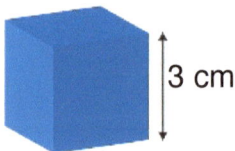

3 cm

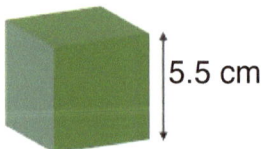

5.5 cm

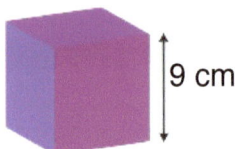

9 cm

The Net of a Rectangular Prism

Study the illustrations below to discover how to draw the net of a rectangular prism.

Surface area of a rectangular prism.

Draw a net for this figure.

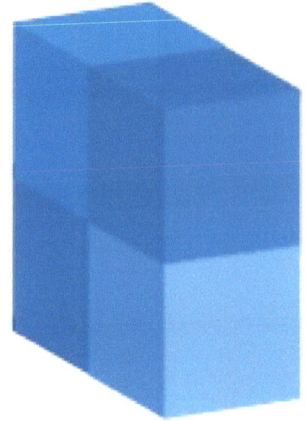

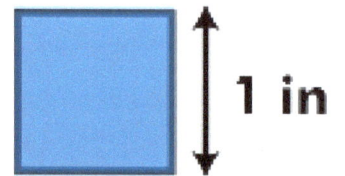

1 in

Calculate the surface area. _____

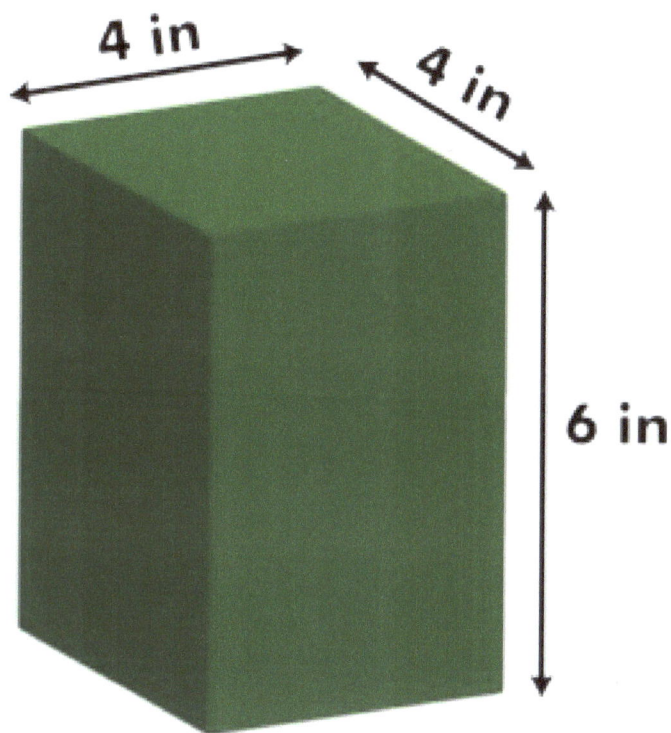

Find the surface area for this rectangular prism using what you have learned.

4 in

4 in

6 in

Name_____

Nets and Surface Area Quiz

1 True or false? Figure 1 is the net of a rectangular prism.

Figure 1

2 The length of the side of a cube is 5 cm. The surface area of the cube is:

A 25 cm^2

B 125 cm^2

C 625 cm^2

D 150 cm^2

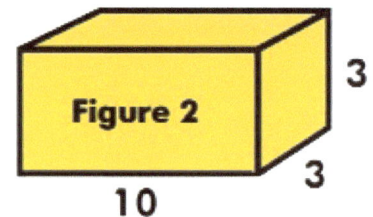

Figure 2 — 10, 3, 3

3 What is the surface area of Figure 2 in square units?

www.ingramcontent.com/pod-product-compliance
Lightning Source LLC
Chambersburg PA
CBHW052053190326

41519CB00002BA/212